Lucy's Engineering Adventure

Written by
Danielle Passaglia

Illustrated by
Gabriella Vagnoli

ISBN 978-1-947192-87-4 (paperback)
ISBN 978-1-947192-88-1 (PDF)
ISBN 978-1-947192-92-8 (eBook)

© 2021 ASHRAE
180 Technology Parkway
Peachtree Corners, GA 30092
www.ashrae.org
All rights reserved.

Text font is Mouse Memoirs.

Library of Congress Control Number: 2021941257

To my Mom and Dad, for giving me the love of numbers and rhyme. – D.P.

To Edward and Luca, the best works of art I have ever made. – G.V.

Thanks to ASHRAE members and staff for the wonderful support. – D.P.

Lucy's Engineering Adventure

"Lucy, my dear, today's the day!
You're coming with me to work—no delay!"

Lucy woke right up and rose with a start and jumped out of bed as fast as a dart.

Filled with excitement, she ran down the hall, jumped over her puppy so she would not fall.

Just like her dad, she'd be an engineer
and learn all about how buildings get here.
How they work and how they are made,
what makes them cozy and how much they weigh.

Dad designs buildings and makes sure they run.

Each workday is different, but that's what makes it fun!

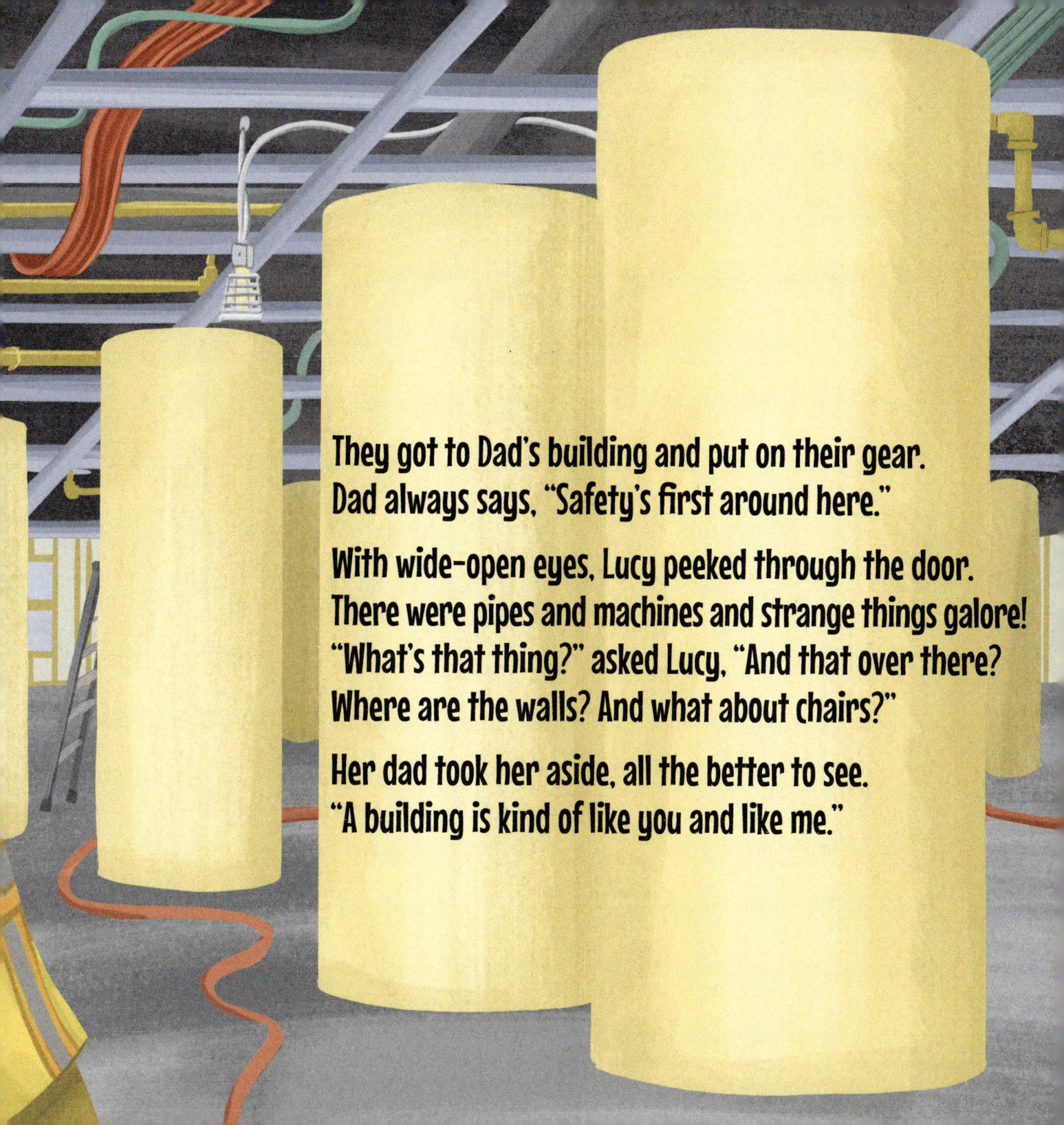

They got to Dad's building and put on their gear.
Dad always says, "Safety's first around here."

With wide-open eyes, Lucy peeked through the door.
There were pipes and machines and strange things galore!
"What's that thing?" asked Lucy, "And that over there?
Where are the walls? And what about chairs?"

Her dad took her aside, all the better to see.
"A building is kind of like you and like me."

"These beams and columns act just like our bones.

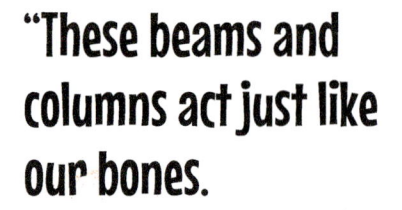

They give us strength so we can stand on our own."

"We need to make sure they are built true and strong,
so people stay safe, and nothing goes wrong."

"Those wires, like our brain, they play a big part.
They bring electricity in, for a start.
They help keep the building brilliant and bright,
by powering things up, like our plugs and the lights."

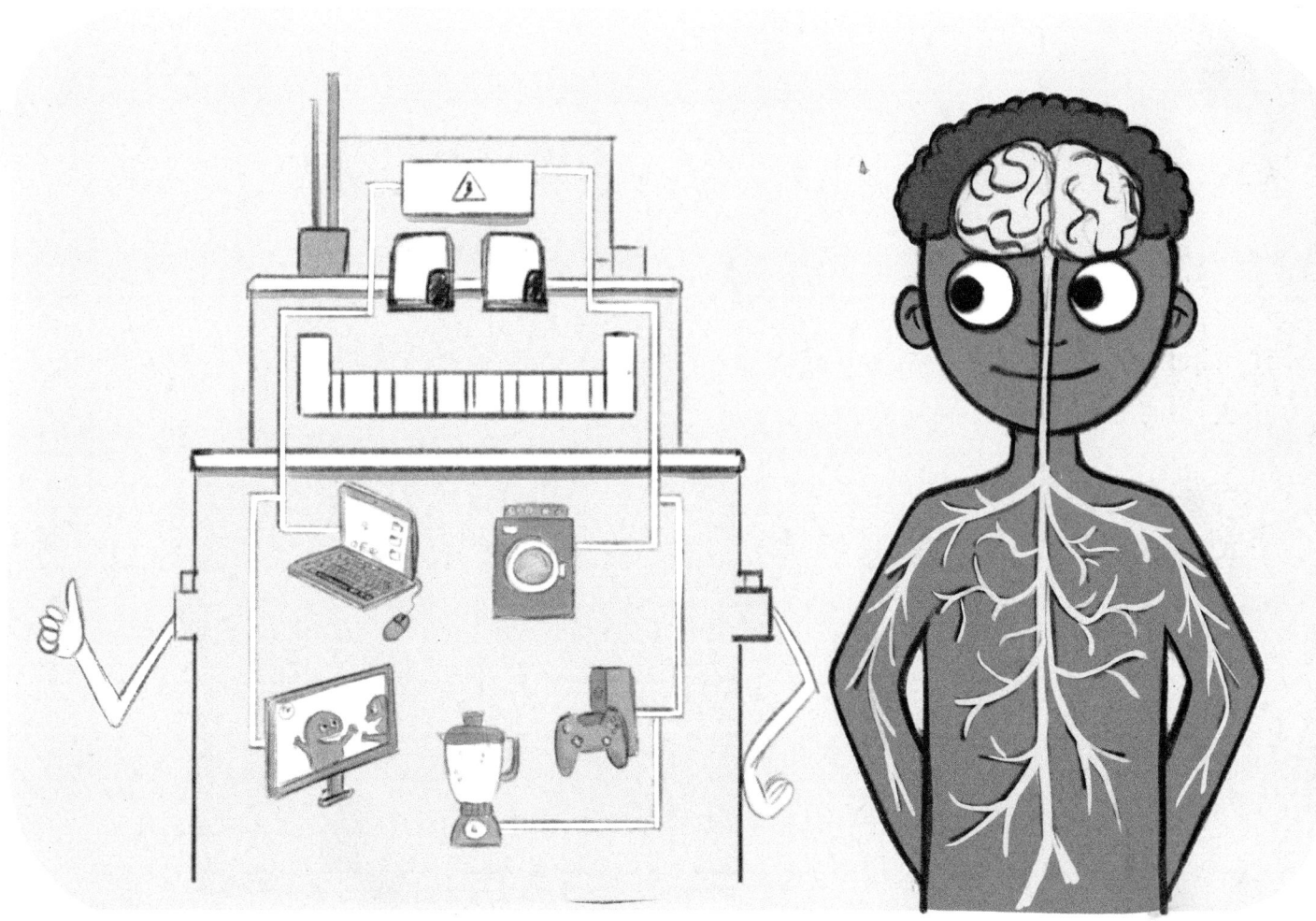

Lucy nodded and, soaking it in,
she slowly broke into a wide, happy grin.
"This is all pretty cool, Dad, but up there, what's that?"
Hanging above was a thing, long and flat.

"That's a duct line, and it is all hollow inside.
Ducts are all through the building, some skinny, some wide.
See—just like our lungs, they provide us fresh air.
Important, though we are not always aware!"

"Ducts carry air we can make warm or make cool,
but doing that sometimes can use lots of fuel,
and using too much may cause air pollution.
So my job is trying to find the solution."

Then Lucy asked, "Dad, tell me—how do you know
what's just enough fuel to make it all go?"

Dad said, "We think hard and we follow some rules
and we work with numbers like you learn in school.
If we do our best and we build it just right,
we're using less fuel by day or by night.
And that helps us keep all the air we breathe clean,
and helps to make all the world healthy and green."

She saw so many things throughout the day, and yet her dream still seemed far, far away. "Dad, all of this thinking seems so hard to do. Tell me—how can I be an engineer, too?"

Lucy's dad was quick to reply.
"Why don't you give this riddle a try?"

"When I walk in a room, it's much too cold.
My breath is showing; it's like the North Pole!
How would you fix this? What do you suggest?
What is the answer you think is the best?"

Lucy thought, and thought, and thought some more. She had many answers, but which to explore?

"What type of room is it? Is it cold outside? What are you wearing? What have you tried? There are lots of answers, but first thing I'd do is close any doors, and I'd close windows, too.

Grab a warm blanket and make a big fire, and maybe for once turn the thermostat higher?"

Lucy's dad grinned, then he hooted and hollered,
"Oh my girl, now you're thinking just like your father!
You already have the right spirit and zest.
Keep asking those questions and trying your best.

Ask about anything that
comes to your mind—
that shows you are curious,
and soon you will find
you are able to change
the world, my dear.
That's how I know you'll
make a great engineer."

This publication was prepared under the auspices of ASHRAE's Student Activities Committee.

Danielle Passaglia, EIT, LEED® Green Associate, Associate Member ASHRAE, is a Mechanical Engineer at Arup in Chicago, IL. She has nearly a decade of experience volunteering with STEM outreach in the K-12 arena via Engineering Ambassadors, TechGirlz, and ASHRAE's Student Activities Committee.

Gabriella Vagnoli has illustrated multiple books, including three books in the Cayuga Island Kids chapter book series by Judy Bradbury (City of Light Publishing). She is Italian and Brazilian, was born and raised in Italy, and currently resides in Illinois with her American husband, two wonderful multicultural kids, and four crazy cats.

Janice K. Means, PE, LEED® AP, FESD, Fellow ASHRAE, Life Member ASHRAE, Professor Emerita at Lawrence Technological University, provided educational oversight for this book. She holds a BA in Secondary Education (Social Science, Physics Minor), a BS in Engineering, and an MS in Mechanical Engineering.

Founded in 1894, ASHRAE is a global professional society committed to serve humanity by advancing the arts and sciences of heating, ventilation, air conditioning, refrigeration and their allied fields. With world headquarters in metro Atlanta, GA, the Society provides essential resources to improve building systems, energy efficiency, indoor air quality, refrigeration and resilience.

ASHRAE's Student Activities Committee develops a comprehensive program for the educational community for the purpose of promoting and encouraging engineering and HVAC&R careers. It administers and promotes student activities at all levels in the educational system. ASHRAE's K-12/STEM Subcommittee works to advance the awareness, understanding, and appreciation of science, technology, engineering, and math (STEM) in the K-12 student population.

ASHRAE is a registered trademark in the U.S. Patent and Trademark Office, owned by the American Society of Heating, Refrigerating and Air-Conditioning Engineers, Inc.

No part of this publication may be reproduced without permission in writing from ASHRAE, except by a reviewer who may quote brief passages or reproduce illustrations in a review with appropriate credit, nor may any part of this publication be reproduced, stored in a retrieval system, or transmitted in any way or by any means—electronic, photocopying, recording, or other—without permission in writing from ASHRAE. Requests for permission should be submitted at www.ashrae.org/permissions.